To our readers, those students just beginning a life of adventure and learning, always remember:

You are braver than you believe, stronger than you seem, and smarter than you think.
~
A.A. Milne

Sounds of the Sea:
A Child's Interactive Book of Fun & Learning

Written & Created by

Brent A. Ford

© 2019 by nVizn Ideas LLC

ISBN 978-1-947348-77-6

www.nviznideas.com

Interactive Components

This version of **Sounds of the Sea** combines the best of both worlds. It is a physical book where children can turn the pages, gaze at the photographs and sit close to a parent or loved-one. It is also a book featuring tech-based, interactive components to extend the fun and the learning.

To access the web-based features, use a mobile device (phone or tablet) with nVizn's QR Code Reader. Our app is FREE and does NOT include advertising or in-app purchases. Our system is also designed to be "kid friendly" - meaning that the app does not open up the entire Internet to young children. Our app only reads codes created by us, so users can access only web content that we create and maintain. Look for the nVizn QR Code Reader in your app store.

After downloading the code reader, simply open the app on your phone or tablet, point it at any one of the many codes throughout the book and you are off. The code reader will automatically take you to a webpage for some learning and fun! (You will need an Internet connection to access these features.)

Try it now....open your QR Code reader and point it at this code.

Think about the sea.
Think about the sounds of the sea.

Did this one come to mind?

Did this sound come to mind?

There are likely other sounds of the sea that did not come to mind. Let's listen.

What is making this sound?

It is a humpback whale singing to other whales. Whale song!

What is making this sound?

It is a whale slapping its tail on the surface of the sea. I wonder why?

What is making this sound?

It is a group of killer whales calling out to each other. I wonder what they are saying.

What is making this sound?

Those are dolphins clicking, squeaking, and whistling to each other. Interesting!

What is making this sound?

That is a seagull calling out to others.
Do you hear other seagulls calling back?

What is making this sound?

That is a manatee gliding through the warm, shallow waters. How about a snack?

What is making this sound?

Those are sea lions lounging on a rock and talking to one another. What about?

Those are harbor seals.
What might they be saying?

What is making this sound?

That is a sea urchin. Really?
Yes, sea urchins can make sounds.

What is making this sound?

That is a cleaner shrimp clicking its claws together to get noticed.

What is making this sound?

That is a clownfish staking out a territory.
Have you heard a fish make a sound?

What is making this sound?

That is a blue-footed booby. Ever seen an animal with blue feet?

What is making this sound?

That is ice cracking and falling off a glacier into the sea. Look ... an iceberg.

What is making this sound?

That is a walrus. It has a very deep voice, don't you think?

What is making this sound?

Those are penguins. Do you think they are talking about how cold the water is?

Look up at the stars, not down at your feet.

Try to make sense of what you see ...

... and wonder about what makes the universe exist.

Stephen Hawking

Now when you think about the sounds of the sea, is this all you will think about?

Learn with Simon

Hi, my name is Simon.

Never seen an animal like me? I am an indri (pronounced IN dree) and I live in a place called Madagascar. To learn more about indris and Madagascar, follow this QR code.

I'm nVizn's mascot - the nVizn indri - and I will be your guide as we learn about the world in which we all live. We'll watch a video or two, do an activity or two, and learn to think and work like a scientist. I'll help get you started on a life-long process of learning about how our world works AND why it works as it does.

So, let's get started!

Sounds of the Sea

From whales that sing to shrimp that click and fish that drum, Earth's ocean is filled with sound. The variation in the sounds created is truly awe-inspiring.

While there are many different kinds of sounds, we should remember (or learn) that sound is created when something vibrates (moves back and forth really fast). When we yell or sing or hum, there is a part of our bodies (a part inside our throats) that we make vibrate. How do we know that? Put your hand on your throat....blow out without making a noise. Now put your hand on your throat and sing, hum, yell. Is there a difference? Sure there is...you do not feel anything when you just blow out and you feel something vibrating in your throat when you make a sound. You are causing that vibration, which we hear as singing or yelling or humming.

If we can make sound by causing our throats to vibrate, how do other animals, including those that live in the ocean, make sounds? Follow this QR code to learn more.

Alike & Different

There are so many different sounds created in Earth's ocean. If we pay attention like a scientist, we can begin to recognize patterns in the kinds of sounds made by particular kinds of animals. Some sounds are very much like other sounds; some sounds are very different. Follow this QR code to look for patterns in the sounds of the sea.

The Bottom of the Sea

Look around Earth and we see some pretty interesting things. We see mountains and cliffs; we see prairies and farmland; we see rolling hills, sand dunes and beaches. But when we look at the ocean, what do we see? That's right, we see water and maybe waves along the beach. But one part of the ocean looks pretty much the same as another part.

But what about the bottom of the ocean? Is it the same everywhere - just flat and sandy? Follow this QR code to investigate.

Sailing the Seven Seas

Poems and songs sometimes mention the seven seas. And perhaps in school, you have learned the names of seven oceans commonly listed as Earth's oceans. And while having a name, like the Atlantic Ocean or the Pacific Ocean, may help us locate a particular area of Earth, such naming also conveys a sense of separateness; that the Pacific Ocean is somehow different than the Atlantic Ocean.

But are the seven named oceans really different?

nVizn Ideas

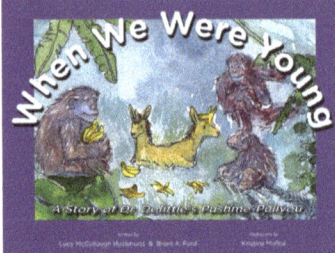

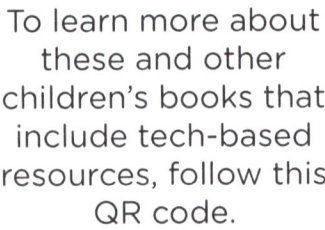

To learn more about these and other children's books that include tech-based resources, follow this QR code.

www.ingramcontent.com/pod-product-compliance
Lightning Source LLC
Chambersburg PA
CBHW040223040426
42333CB00051B/3430